I0409279

HRH Prince John Charles Wright

HRH Prince Joe Duncan Wright

HRH Prince John Richard Wright

Steel manufacturing in Switzerland

1st Published in 1999

2nd Publication in 2008

Table of Contents

Steel manufacturing in Switzerland

Switzerland is not a major steel-producing country, and its steel manufacturing industry is relatively small compared to other global players. However, it does have a presence in the steel industry with some notable companies and steel-related activities. The country does have a strong automotive sector that relies on steel and other materials for vehicle production and manufacturing.

Switzerland does have a significant presence in the automotive industry, particularly in the luxury and high-end segment. The country is known for its precision engineering, and Swiss companies contribute to the automotive sector by manufacturing high-precision components and parts used in vehicles.

Key points

Switzerland's steel production is limited compared to larger steel-producing countries. The country mainly focuses on specialty steels and high-value steel products. Switzerland is known for producing high-quality specialty steels used in various industries, including automotive, aerospace, machinery, and medical equipment.

These steels often require precise engineering and have specific applications. While steel production is not the main focus, Switzerland is an important hub for steel trading and processing. It imports steel products from other countries and processes them for specific applications. Switzerland invests in research and development related to steel technologies, innovation, and advanced manufacturing processes.

Swiss companies are known for their expertise in metalworking and precision engineering. Switzerland invests in research and development related to automotive technologies, including materials and manufacturing processes. Collaborations between Swiss research institutions and international automotive companies contribute to innovation in the sector. Switzerland places a strong emphasis on sustainability and environmental protection. Companies in the steel manufacturing sector adhere to strict environmental regulations and adopt eco-friendly practices.

The steel industry globally faces challenges such as fluctuating demand, competitive pressures, and trade dynamics. Switzerland's steel manufacturers must navigate these challenges while focusing on value-added products and services. Steel manufacturing in Switzerland is often integrated with other industries, such as engineering, automotive, and machinery, which benefit from the country's expertise in precision manufacturing.

Vehicle manufacturing in Switzerland

Switzerland does not have domestic automobile manufacturers and automobile manufacturing facilities on large-scale of its own. Most vehicles used in Switzerland are imported from other countries, primarily from Germany, France, Italy, and Japan. The automotive industry in Switzerland relies on importing steel and other materials from various countries to meet its manufacturing needs.

Car manufacturers and suppliers are increasingly seeking to use lightweight and sustainable materials, which could influence the demand for advanced high-strength steel and other materials. Switzerland is known for its luxury and high-end automotive market. Switzerland's automotive market focuses on luxury and high-end vehicles, where steel is used for various components, structural parts, and chassis.

The country has a significant demand for premium vehicles, and these vehicles often incorporate high-quality steel in their construction. Switzerland invests in research and development related to automotive technologies, including materials and manufacturing processes. Collaborations between Swiss research institutions and international automotive companies contribute to innovation in the sector. Swiss companies are renowned for their expertise in precision engineering and manufacturing.

Swiss companies are renowned for their expertise in precision engineering and manufacturing. They often produce high-quality components that contribute to the overall performance and safety of vehicles. Similar to many developed countries, Switzerland's automotive industry places a strong emphasis on environmental considerations and sustainability.

Car manufacturers and suppliers are increasingly seeking to use lightweight and sustainable materials, which could influence the demand for advanced high-strength steel and other materials. While they may not produce steel on a large scale, they contribute to the automotive industry by manufacturing high-precision components and parts used in vehicles. Switzerland invests in research and development related to automotive technologies, including materials and manufacturing processes.

Collaborations between Swiss research institutions and international automotive companies contribute to innovation in the sector. Like many developed countries, Switzerland's automotive industry places a

strong emphasis on environmental considerations and sustainability.

Car manufacturers and suppliers are increasingly seeking to use

lightweight and sustainable materials, including advanced high-

strength steel and aluminum alloys.

Aerospace industry in Switzerland

Switzerland has a well-established aerospace sector that includes the manufacturing of aircraft components, systems, and technologies. The country is known for its expertise in precision engineering, which contributes to the aerospace industry's success. Similar to other industries, Switzerland may import steel and other materials required for aerospace manufacturing from various countries.

Swiss companies play a vital role in precision engineering for the aerospace industry. They manufacture critical components and parts with high precision, meeting the stringent requirements of the aviation sector. Switzerland invests in research and development related to aerospace technologies and materials.

Collaborations between Swiss research institutions and aerospace companies contribute to innovation in the sector. The aerospace industry, including aircraft manufacturing, is increasingly focused on sustainability and eco-friendly practices. Companies are exploring lightweight materials, including advanced steel alloys, to improve fuel efficiency and reduce environmental impact. Beyond manufacturing, Switzerland also provides aerospace-related services, such as maintenance, repair, and overhaul (MRO) operations, for both domestic and international aircraft. While Switzerland may not be a primary steel manufacturer for the aerospace industry, its contribution lies in precision engineering, research, and value-added services within the aerospace supply chain.

Machinery manufacturing in Switzerland

Switzerland has a well-established machinery manufacturing sector

that produces a wide range of machinery and equipment, including

industrial machinery, machine tools, automation systems, and more.

Steel is a fundamental material used in the machinery industry due to

its strength, durability, and versatility.

It is used in manufacturing machinery components, such as gears,

shafts, frames, and structural supports. Swiss companies are renowned

for their expertise in precision engineering, which is vital for producing

high-quality machinery and mechanical components. Steel is often

machined to precise tolerances to ensure optimal performance.

Switzerland is also known for producing custom-made machinery and equipment tailored to the specific needs of industries like pharmaceuticals, watchmaking, and precision manufacturing.

While Switzerland may import some steel for its machinery manufacturing needs, it also relies on local suppliers and steel manufacturers within Europe. Swiss machinery manufacturers invest in research and development to continuously improve their products and stay at the forefront of technological advancements. Like in other industries, the machinery sector in Switzerland is increasingly considering sustainable practices and environmental impact. Manufacturers may explore eco-friendly materials and energy-efficient technologies in machinery production.

Advancements

Switzerland is known for its precision engineering and technological advancements in various industries, including aerospace, machine manufacturing, and vehicle manufacturing. Swiss companies are involved in the development of advanced aerospace technologies, including aircraft components, avionics, and propulsion systems. They contribute to the design and manufacturing of high-tech aerospace products.

Switzerland has been at the forefront of the drone and unmanned aerial vehicle (UAV) industry. Swiss companies have developed drones for various applications, including aerial photography, surveying, agriculture, and surveillance. Switzerland is actively involved in space technology and satellite development. The country has contributed components and instruments for various space missions, including satellite missions led by the European Space Agency (ESA). Swiss machine manufacturers have embraced automation and robotics to improve productivity and efficiency. Advanced robotics and cobots (collaborative robots) are integrated into manufacturing processes to handle repetitive tasks and assist human workers.

Switzerland has embraced Industry 4.0 principles, which involve the integration of digital technologies into manufacturing processes. This includes the use of data analytics, artificial intelligence, and the Internet of Things (IoT) to optimize production and decision-making. Swiss companies have adopted additive manufacturing techniques to produce complex and customized parts with reduced waste and faster lead times.

Swiss machine manufacturers have integrated advanced automation and robotics into their production processes. Collaborative robots and smart automation have improved productivity and allowed for more complex and customized manufacturing. Switzerland has adopted digitalization and Industry 4.0 principles to optimize production and enhance decision-making through data analytics and the Internet of Things (IoT). Swiss companies have embraced additive manufacturing techniques to create intricate and lightweight components, reducing waste and lead times.

Swiss companies have developed miniaturized satellites and components for space missions. These small satellites are cost-effective and can be deployed in constellations to provide various services, such as earth observation and communication. Swiss companies have made significant strides in the development of UAVs for various applications, including mapping, surveillance, agriculture, and environmental monitoring. Switzerland excels in manufacturing precision aircraft components and avionics systems. These components are critical for the performance and safety of aircraft and are in demand globally.

Switzerland has invested in space technology and satellite manufacturing, developing high-quality components and instruments for space missions. The country's expertise in precision engineering and microtechnology has been pivotal in miniaturized satellite development.

Swiss companies have embraced the potential of UAVs and drones, advancing the development of cutting-edge aerial technology for various applications, including mapping, monitoring, and surveillance.

Switzerland is renowned for producing precision aircraft components and avionics systems, contributing to the global aerospace supply chain.

Swiss machine manufacturers are known for producing high-precision machinery, including CNC machine tools, lathes, milling machines, and industrial robots. The focus on precision engineering allows Swiss machinery to be in demand globally. Swiss machine manufacturers excel in producing custom-made machinery and equipment tailored to the specific needs of industries such as watchmaking, pharmaceuticals, and precision manufacturing. Switzerland has embraced automation and Industry 4.0 concepts, integrating smart technologies into machinery to enhance productivity and efficiency.

Switzerland is known for its luxury and high-end automotive market. The country imports and sells luxury vehicles from renowned brands, making it a hub for exclusive automobiles. Like many countries, Switzerland has shown a growing interest in electric mobility. The government and private sector are encouraging the adoption of electric vehicles, leading to the development of charging infrastructure and sustainable transportation solutions.

Switzerland has shown a strong commitment to electric mobility, with an increasing number of electric vehicles (EVs) on the roads. The government provides incentives for EV adoption, and charging infrastructure is expanding rapidly. Switzerland has shown strong support for electric mobility, with a focus on increasing the adoption of electric vehicles and expanding charging infrastructure.

Swiss companies are actively engaged in research and development of autonomous driving technologies, contributing to the advancement of driver assistance systems and autonomous vehicles. Switzerland is exploring the potential of hydrogen fuel cell vehicles as a sustainable transportation option. Swiss manufacturers are using lightweight materials like carbon fibre and aluminium to improve vehicle performance and reduce emissions

Swiss companies are actively involved in research and development of autonomous driving technologies. They are collaborating with global players to develop advanced driver assistance systems and autonomous vehicles. Swiss companies are exploring the use of lightweight materials, such as carbon fibre and aluminium, to improve vehicle performance and reduce emissions. Switzerland is also exploring the potential of hydrogen fuel cell vehicles as a sustainable alternative to conventional internal combustion engines.

To foster further growth in these industries, Switzerland continues to invest in research and development, promote collaboration between industry and academia, and support startups and innovative projects. Additionally, the country's commitment to sustainability and eco-friendly practices aligns with global trends, positioning Swiss industries for continued success in the future.

Swiss companies are involved in research and development of automotive technologies, including autonomous driving, connected vehicles, and lightweight materials to improve vehicle performance and reduce emissions. Switzerland has hosted Formula E races, showcasing the country's support for electric racing and clean energy initiatives.

To enhance capacity and capability, Switzerland encourages collaboration between academia and industry, fosters research and development, and supports startups and innovative projects. The government's investment in research and education also plays a crucial role in nurturing talent and expertise in these industries.

Switzerland's commitment to sustainability and environmental responsibility aligns with global trends, positioning its aerospace, machine manufacturing, and vehicle manufacturing sectors for continued growth and success in the future.

Switzerland has made significant progress in promoting the use of renewable energy sources, such as hydropower, solar, and wind energy. Many manufacturing facilities in these industries are transitioning to renewable energy, reducing their carbon footprint and environmental impact. Swiss manufacturers in aerospace, machine manufacturing, and vehicle manufacturing sectors are increasingly adopting energy-efficient processes and technologies to minimize resource consumption and waste generation.

The industries are exploring the use of sustainable materials in their products and manufacturing processes. For instance, the adoption of lightweight and eco-friendly materials in aerospace and vehicle manufacturing reduces fuel consumption and emissions. Switzerland promotes circular economy principles, encouraging industries to design products that can be easily recycled or refurbished, thus reducing waste and extending product lifecycles.

The Swiss government has set ambitious emission reduction targets, which have encouraged companies in these sectors to implement strategies to reduce greenhouse gas emissions and contribute to climate action. Switzerland invests in research and development of sustainable technologies in these industries. This includes advancements in electric mobility, clean energy technologies, and eco-friendly manufacturing processes.

Many companies in the aerospace, machine manufacturing, and vehicle manufacturing sectors have embraced CSR initiatives, focusing on social and environmental responsibilities, including sustainable supply chain management. The Swiss government collaborates with private industries, research institutions, and non-governmental organizations to address environmental challenges collectively. This collaboration facilitates the adoption of sustainable practices and the development of eco-friendly technologies.

By aligning with global trends towards sustainability and environmental responsibility, Switzerland's aerospace, machine manufacturing, and vehicle manufacturing sectors are well-positioned to remain competitive and contribute positively to the global effort in combating climate change and achieving a greener future. This commitment to sustainability not only strengthens the industries' resilience but also enhances their reputation as responsible global players in the manufacturing sector. By embracing sustainability and environmental responsibility, Switzerland's industries are not only meeting the demands of a changing market but also contributing to a greener and more sustainable future.